Molecular
Memories

Molecular Memories

Robert G. Jahn
and
Brenda J. Dunne

ICRL Press
Princeton, New Jersey

Molecular Memories
By Robert G. Jahn and Brenda J. Dunne
Copyright © 2015 by Robert G. Jahn and Brenda J. Dunne

ISBN: 978-1-936033-21-8

Cover Image: Jag_cz/BigStock

Table of Contents

Introduction

FROM 1979 TO 2007, an extraordinary enterprise thrived in what was previously a small storage area next to the machine shop in the basement of Princeton University's School of Engineering and Applied Science. Known as the Princeton Engineering Anomalies Research (PEAR) laboratory, the program was staffed by a small interdisciplinary team of engineers, psychologists, and physicists, whose primary purpose was the study of consciousness-related anomalies and their implications for engineering technology.

What made PEAR unique was its integration of rigorous scientific methodology with diverse spiritual traditions, sophisticated analyses with intuitive aesthetics, and state-of-the-art technology with warm hospitality—all reflecting the complementary strengths, styles, and personalities of its two founders. As evidence of the productivity engendered by this synergy, over the years the laboratory published several hundred scholarly articles and reports and attracted countless professional and personal visitors from around the world, and it rapidly acquired a reputation as a leading program in its field.

Concomitant with its nearly 40 years of basic research activity and its numerous scientific articles and presentations, the authors heretofore have published a trilogy of archival textbooks that have endeavored to chronicle comprehensively the purposes, accomplishments, and deductions of the specific research projects they have undertaken:

Margins of Reality: The Role of Consciousness in the Physical World (2009): a survey of PEAR's preliminary studies and the relevance of this research within a historical framework and a variety of scholarly perspectives; [1]

Consciousness and the Source of Reality: The PEAR Odyssey (2011): a comprehensive review of the PEAR empirical results and conceptual models; [2]

Quirks of the Quantum Mind (2011): a metaphorical exploration of the role of consciousness in quantum mechanics, along with its pertinent metaphysical implications and applications, with an extensive Appendix of relevant quotations from the patriarchs of quantum mechanics. [3]

---

1 R.G. Jahn and B.J. Dunne (2009). *Margins of Reality: The Role of Consciousness in the Physical World.* Princeton, NJ: The ICRL Press. (Originally published in 1987 by Harcourt Brace Jovanovich.)

2 R.G. Jahn and B.J. Dunne (2011). *Consciousness and the Source of Reality: The PEAR Odyssey.* Princeton, NJ: The ICRL Press.

3 R.G. Jahn and B.J. Dunne (2011). *Quirks of the Quantum Mind.* Princeton, NJ: The ICRL Press.

While this documentation has afforded an inclusive representation of the technical and philosophical aspects of the research, there yet remain what we regard as two important, even critical, aspects of the PEAR program that are at least as essential as its methodology and data analyses for any full comprehension of the nature of this enterprise. One is the necessity of incorporating certain essential *subjective* elements into any *scientific* studies of consciousness, an issue that has been developed at length in an article titled "Science of the Subjective" [4] and summarized as a chapter in *Consciousness and the Source of Reality: The PEAR Odyssey.*

The second feature, which addresses the crucial role of the interpersonal bond between PEAR's founding partners, is more elusive to relate and sensitive to specify in detail. Yet it represents a critical component of the program, without which any portrayal of it would be incomplete. Hence, this volume will attempt to address that dimension, albeit somewhat implicitly.

The two individuals who comprised this bond approached the tasks before them from vastly different personal backgrounds, experiences, and styles, and originally had little in common beyond a shared vision and a determination to bring that vision into being. It is virtually impossible to capture the ineffable dynamics of this remarkable interpersonal liaison, which transcends

---

4 R. G. Jahn and B. J. Dunne (1997). "Science of the Subjective." *Journal of Scientific Exploration*, 11 (2) pp. 201-224.

any ordinary friendship, in any linear descriptive narrative. Instead, this book takes the approach of presenting a number of indicative episodes, or "vignettes," experiences that they shared over the evolution of the research program. Their "molecular bond," and its associated complementarity, provided the foundation for the creative PEAR experience. No one of these stories alone can fully encapsulate the extraordinary subtleties of their respective individual contributions. But, just as in a physical molecule, it is only in the interaction of the constituent "atoms" that the characteristics of the unified system become apparent, and this personal and professional bond embodies the heart of the PEAR enterprise.

These episodes may appear rather polyglot: some address humorous or light-hearted events, while others broach profoundly serious issues; some deal with prosaic matters, and others touch on less tangible metaphysical aspects. Taken collectively, however, this assortment of vignettes may at least provide a glimpse of the magic that can arise from mutual respect, trust, love, and openness to uncertainty in a profound relationship. It is not an exaggeration to say that the authors have come to believe that they have been dealing here with a fundamental aspect of life and reality, which their empirical research has verified and developed. The only coda we would add to endorse the inclusion of this dimension into the creative mix of any scholarly study, or indeed of any creative enterprise, is our personal assurance of its ineffable effectiveness.

1. The Atoms

BOB JAHN was born in central New Jersey in 1930 to loving and rather conservative Presbyterian parents, but he spent most of his early years in Wilmington, Delaware. As an only child, he frequently amused himself by inventing games with complex rules, many of them associated with baseball or other major sports, passions of his for as long as he can remember. He was an outstanding student whose exceptional intellectual and creative skills earned him a scholarship to Tower Hill School, a distinguished Wilmington preparatory school ranked among the top 50 private day schools in the U.S. In 1947 he won one of two Pepsi Cola National Scholarships in the state of Delaware, which enabled him to attend Princeton University, from where he graduated in 1951, *summa cum laude*, with a joint degree in mechanical engineering and physics. He remained at Princeton to earn his PhD in physics.

Most of Bob's professional career followed a relatively traditional, albeit highly successful, academic trajectory. After several years teaching at Lehigh University and the California Institute of Technology, he returned to Princeton in 1962 with his growing family to accept an appointment as Assistant Professor of Aerospace Sciences. There he established a major program in electric space propulsion that soon became recognized worldwide as one of the foremost laboratories in its field, and where a long sequence of future space scientists received their professional training under his supervision. His seminal textbook, *Physics of Electric Propulsion*, was published in 1968 and has twice been reissued. Bob was promoted to Associate Professor in 1964, to Full Professor in 1967, and in 1971 he was appointed Dean of Princeton's School of Engineering and Applied Science, a position he held for 15 years.

Beyond his extensive administrative and teaching responsibilities, Bob has also been a Fellow of the American Physical Society and of the American Institute of Aeronautics and Astronautics. He has served as chairman of the Board of Trustees of Associated Universities, Inc., as chairman of the AIAA Electric Propulsion Technical Committee, as associate editor of the *AIAA Journal*, and as a member of the NASA Space Science and Technology Advisory Committee. He has also been a member of the Boards of Directors of Hercules, Inc.; Drexel University; and Roy F. Weston, Inc. In 1969 he was awarded the Curtis W. McGraw Research Award of the American Society of Engineering Education, and in 1986 he received an honorary Doctor of Science degree from Andhra University in India. In 2012 he was awarded the prestigious AIAA Wyld Propulsion Award. Bob was also one of the founders of the Society for Scientific Exploration.

His outstanding academic and scholarly credentials were beyond question, and Bob was highly regarded by his colleagues and students. But unknown to most of them, he has maintained a lifelong personal interest in spirituality and the nature of consciousness.

It wasn't until 1977, however, when he had occasion to supervise an independent student project in human/machine anomalies, that he was able to forge a path that would enable him to integrate these interests with his dedication to science and engineering. The student in question had read of the groundbreaking research of Helmut Schmidt, a physicist who had demonstrated a correlation between human intention and the behavior of a quantum mechanical random number generator,[5] and she proposed to design and construct an electronic random number generator to determine whether Schmidt's results could be replicated. When other members of her departmental faculty expressed no interest in the project, Bob offered to supervise it himself. The positive results of this modest pilot experiment were sufficiently intriguing that he felt it was worth pursuing more systematically.

The possibility that human consciousness might influence the behavior of random physical systems had profound personal and professional implications for him, as well as for contemporary engineering technology. With the encouragement and financial support of James S. McDonnell, a prominent aerospace magnate and Princeton alumnus who shared Bob's interest in this topic, he undertook the establishment

---

5 Schmidt, H. (1967). "New correlation between a human subject and a quantum mechanical random number generator," Boeing Scientific Research Laboratories Document D1-82-0684.

of a research program in engineering anomalies in his engineering school. Needless to say, this did not endear him to the university's administration, nor to many of his academic colleagues, who regarded this enterprise as inappropriate and as an embarrassment to the institution. Nonetheless, Bob persevered in

Bob

his belief in the importance of such research and, by judicious invocation of the principle of academic freedom, succeeded in overcoming this resistance and establishing the PEAR laboratory in 1979.

Given his numerous responsibilities as Dean and as Director of his Electric Propulsion program, he realized that he would need a colleague who could handle the day-to-day activities of this new laboratory, and he set out to identify an appropriate person for this position. The options were severely limited, since he was searching for an individual with scientific credentials, one who had some knowledge and experience of psychic research, and, in particular, a person who would provide a complement to his own skills and background, preferably a woman. He began his search by attending meetings of the Parapsychological Association to familiarize himself with the researchers in that field and the work that was being pursued therein.

BRENDA DUNNE's background was as erratic as Bob's was structured. She was an adopted child born in 1944, who grew up in Brooklyn, New York, with a background characterized by an unstable family environment, volatile relationships, and diverse religious traditions. At a very young age she learned that the various rule systems to which she was being subjected were situation-specific and mutable, and she became adept at adjusting to unpredictable circumstances with a certain degree of flexibility. Her refuge from the uncertainty of her surroundings was an early fascination with fairy tales, which led to the study of world mythologies, and ultimately to a lifelong interest in the history of religion and the spiritual and mystical traditions of different cultures.

An early marriage and ensuing motherhood precluded her pursuit of a higher academic education until several years later. Only after her family moved to the Chicago area and her two children were in school full time did she enroll at Mundelein College, where she earned joint undergraduate degrees in psychology and the humanities *magna cum laude*, and completed two independent honors theses: one in experimental parapsychology, and one in altered states of consciousness. In 1976 she was accepted into a PhD program at the University of Chicago's Committee on Human Development.

In addition to her academic studies, Brenda was active in local politics, organized several of her community's annual 4th of July parades, practiced yoga and various forms of meditation, and developed a strong interest in environmental concerns. These intellectual and pragmatic experiences eventually made her realize that the life of a suburban housewife was incompatible with her personal interests and aspirations, and she was divorced in 1977.

An elective class in experimental parapsychology at Mundelein College led to an independent research project in which she carried out the first successful replication of the celebrated remote viewing studies conducted at Stanford Research Institute that had just recently been reported,[6] and she gave a presentation on this initial experiment at a meeting of the Parapsychological Association in 1977. Several further such formal investigations in the Chicago area, together with an ongoing series of inexplicable personal experiences, persuaded her that such anomalous phenomena were not only real, but potentially fundamental for any full understanding of human consciousness and the nature of reality, even though they were not currently respected by mainstream science.

Her background in diverse spiritual and mystical traditions, her interests in the philosophy of mind, and her studies of altered states of consciousness led her to consider the possibility that the materialistic rule system that guided the prevailing scientific models might not be the only viable one of several alternative modes

6 Puthoff, H.E. and Targ, R. (1976). "A perceptual channel for information transfer over kilometer distances: Historical perspectives and recent research." *Proceedings of the IEEE*, Vol. 64, No. 3.

of representation. She had a profound respect for the scientific method and was persuaded that it could reasonably be applied to investigating alternative models that would incorporate consciousness and the subjective aspects of experience. Yet she knew that her own limited scien-

Brenda

tific training was inadequate to pursue this idea in any depth, and during the next several years she attempted to remedy this via her graduate studies and extensive independent reading.

In 1977 she had a life-changing experience while attending a lecture by Fritjof Capra, who was discussing his recently published book *The Tao of Physics*.[7] In the course of his presentation about quantum mechanics, a topic with which Brenda had no familiarity, Capra projected a representation of Shiva Nataraja, the paradoxical Hindu god who represents the eternal cycle of creation and destruction, superimposed on an image of a cloud chamber, a device physicists use to detect and study elementary particles. This striking symbolism, which incorporated both the metaphysical and scientific conceptions of reality,

---

7 Capra, F. (1975). *The Tao of Physics: An Exploration of the Parallels Between Modern Physics and Eastern Mysticism*. Boston: Shambhala Publications.

seemed to offer a path by which she might be able to integrate this science with her other interests, and she became determined to learn more about quantum mechanics and what it might reveal about the nature of consciousness.

2. The Molecule

ALTHOUGH BOB had heard Brenda give a presentation in 1977, they didn't actually meet each other until the following year. It was in August 1978, during a coffee break of the annual meeting of the Parapsychological Association at Washington University in St. Louis, Missouri. Brenda had just given a brief talk describing a series of experiments in long-distance remote viewing, and Bob was at the meeting to acquaint himself with the players in the field in the hope of identifying a candidate who could manage the day-to-day operations of the program he was planning to establish at Princeton. Having been informed that he had asked to meet her, she approached him as he was surrounded by a group of attendees.

"I understand you were looking for me," she said.

"You might say," he replied, looking at her intently with his deep brown eyes. "I enjoyed your talk. What are you planning to do next?"

"I'm concerned about the reliability of the analytical techniques we've been using to evaluate the experimental results because I see a possibility of subjective bias in the judging procedure," she replied.

"What do you have in mind?" he asked.

"I'm not quite sure, but it seems that the data should be evaluated by a more objective technique that would eliminate the potential subjectivity of the human judge," she answered.

"Have you considered using a standardized list of descriptors?" he suggested.

"What an excellent idea!" she responded.

And within the next 10 minutes they devised a new judging procedure that would solve the problem effectively and become one of the foundation stones of the future PEAR research program.

She was struck by his ability to get to the core of a problem so quickly. She was also taken by his red baseball cap and warm smile, and the playful boy she perceived hiding behind the intimidating façade of a Dean of Engineering from a prestigious university. He was impressed by her ability to identify a design flaw and to recognize an alternate approach so quickly.

"I'm thinking of starting a research program at Princeton and am looking for someone who might be capable of working with me as a laboratory manager. Would you be interested in the job?"

"You bet!" she responded with what might have been an excessive show of enthusiasm.

"Let's talk about this further a little later," he said as he turned to speak to another guest. She stood there, annoyed with herself for what she felt was probably an unprofessional reaction.

That's how it started—just like a fairy tale. They couldn't have been more different in backgrounds and styles. Bob was a serious scientist of the highest academic credentials—pragmatic and incisive. He spoke in complete sentences, with no words wasted. Brenda was a recently divorced housewife and graduate student with little formal scientific training—outspoken and impulsive. He was a devoted baseball fan, she disliked sports; she liked rock music, he preferred classical compositions. His food preferences were "meat and potatoes," she preferred ethnic fare. He was a physicist and engineer; she had barely managed to make her way through high school math. Her scholarly interests ran to psychology, philosophy, and the history of religion; his were aerospace science and quantum mechanics.

At first, it looked as though they would not have too much to say to one another. But what they had in common was a shared passion for learning and a profound sense that there was something essential to be learned about the nature of consciousness through

Bob and Brenda, ca. 1982

the study of anomalous phenomena. Both tended to be impatient and highly motivated when addressing a new idea, and capable of focusing intently on the problem at hand, although they frequently approached it from different perspectives. Most importantly however, they quickly developed a deep trust for each other's styles and found creative solutions that blended their respective insights. In the ensuing years they were seldom at a loss for conversation.

One such conversation that was indicative of the nature of their "molecule" arose in the context of Brenda's concern that she frequently felt out of place in the academic/engineering environment in which she found herself, and that people seemed not to take her seriously. Bob told her that her problem essentially stemmed from the fact that she was a unicorn who had painted herself with black stripes in the hope of passing as a zebra. This ploy worked reasonably well from a distance, he said, but when she was interacting with a herd of zebras it was inevitable that they would recognize that she was not one of their own.

"You need to wash off all that paint and just be yourself," he told her.

"That's easy for you to say" she responded, "but do you realize what happens to unicorns when they're caught?"

"Yes," he replied, "but consider what is required to catch a unicorn!"

A few months after their initial meeting, Bob traveled to Evanston, Illinois, where Brenda was then residing. During a walk along the shore of Lake Michigan, they discussed what might underlie the kind of anomalies they were considering studying. Brenda attempted to describe her impressionistic ideas of how the experiences of a single individual tended to differ from those arising in a shared interaction with another person. Bob responded that her subjective representation bore a certain metaphorical resemblance to the dynamics of the covalent bond in quantum science, and he proceeded to explain the concept to her, drawing an image in the sand. At that point they realized that the question and the answer drew from different domains, in fields that seldom communicated with each other but were essentially complementary. "You physicists already have the answer," Brenda observed, "but you don't know what the question is!"

Over the years they continued to explore the complementarity of what one regarded as the quantum mechanics of physical reality, and the other viewed as the "psychology of atomic particles," in conversations that proved both informative and enjoyable. Nor were the analogies limited to covalent bonds, but extended across a wide spectrum of other quantum mechanical concepts, and even well beyond the confines of physics. They came to refer to Bob as the "particle:" he was a systematic thinker, well oriented in space and time, and inclined to ask, "What is your point?" Brenda, on the other hand, was the "wave" dimension of the molecule: she had a strong intuitive sense and refined multi-tasking skills, and was good at dealing with people. Bob made all the analytical decisions about the research protocols and designs, while Brenda handled the interpersonal interactions with operators and visitors. They soon began to describe their moods and functions in quantum mechanical terms, speaking in terms of their prevailing "quantum numbers," "spins," and "orientations." Playing with such analogies led to the realization that in this metaphorical sense, quantum mechanics offered an apt model of how consciousness organizes and represents its experiences, along with the recognition that human consciousness itself was simply a reflection of Consciousness—the primordial ordering principle of the universe.

3. Quail Roost

IN DECEMBER 1978, when PEAR was still in an embryonic stage, we were invited to join a select gathering of individuals who shared a common interest in the study of consciousness and its personal, societal, and spiritual implications. Through the efforts of U.S. Representative Charlie Rose of North Carolina, who was one of the participants, the meeting convened at a charming North Carolina retreat center called "Quail Roost." This proved to be a propitious setting for such a gathering, replete with natural attractions such as flocks of resident peacocks, herds of deer, many species of birds and other wildlife, comfortable accommodations, and excellent food.

opendurham.org

Quail Roost, Rougemont, NC

In addition to Rep. Rose, an outspoken advocate for research in this field, the attendees included a roster of impressive participants from several other major initiatives. These included John Fetzer, a radio and television pioneer who was perhaps better known as the owner of the Detroit Tigers baseball team, and who had established The Fetzer Institute in 1962 "to foster awareness of the power of love and forgiveness in the emerging global community;" Willis Harman, Judith Skutch, and several others actively involved with the Institute of Noetic Sciences that had been founded in 1973 by astronaut Edgar Mitchell; author Marilyn Ferguson, a founding member of the Association of Humanistic Psychology and editor of the popular science newsletter *Brain/Mind Bulletin*, who was just then in the process of writing the influential book *The Aquarian Conspiracy: Personal and Social Transformation in the 1980's*;[8] and other participants of comparable caliber.

In addition to the benefits of interacting with these interesting people and acquiring new friends and a fresh set of perspectives on the directions in which consciousness research was moving, the event provided an opportunity to announce the formation of our own embryonic academic research program and to benefit from the insights and recommendations that were offered by the other participants.

---

8 Ferguson, M. (1980). *The Aquarian Conspiracy: Personal and Social Transformation in the 1980s*. Los Angeles: Jeremy P. Tarcher.

During dinner one evening, Brenda was seated next to John Fetzer who described his ongoing interest in so-called "paranormal phenomena." He asked her whether she had ever had such an experience, to which she replied in the affirmative.

"Can you make these things happen at will?" he asked.

"Not really," she replied, "but I can sometimes tell when something unusual is about to happen: the edges of reality tend to become sort of fuzzy."

"Well, my dear, if you notice things getting fuzzy be sure to let me know," he requested.

Sometime later in the evening, perhaps aided by a few glasses of excellent wine, Brenda felt her consciousness diffusing and shifting, and she mentioned this to John. He suggested that she try to extinguish one of the candles on the table; the candle remained unaffected by her efforts but instead one of the metal links in her copper necklace had inexplicably unbent and had fallen into her lap!

As a result of John's and Bob's shared passion for baseball and John's fascination with Brenda's views on anomalous experiences, a strong friendship quickly developed among the three of us, and before leaving Quail Roost John invited us to come to Detroit to see a baseball game with him. We

subsequently purchased two toy tigers clad in baseball garb; we sent one to John, which he kept on display in his office, and the other sat perched on Bob's desk for many years.

The following September we accepted John's invitation and attended a Tigers' game with him at Briggs Stadium in Detroit. The Tigers, who were at the bottom of their league at the time, were playing the reigning Baltimore Orioles that day, and a rookie player named Kurt Gibson, who went on to become a World Series hero and future manager of the Detroit team, was making his major league debut. Bob was euphoric, especially when Hall of Famer Al Kaline, then a radio announcer, joined us in the owner's box. Brenda, on the other hand, had little idea of what was going on in the game but felt some compassion for the young rookie, imagining how it must feel for him to be playing his first major league game with the owner looking on.

When John asked her whether the Detroit team would win, she casually replied "Sure, how about five to nothing? Anything else we can do?"

"How about a triple play? They're pretty rare," suggested John with a smile.

"We'll work on it," Brenda responded.

There was no triple play, although there were three double plays in the game, as well as an inordinate number of broken bats. Young Gibson played superbly, and by the middle of the last inning the Tigers had won, 3-0. John was pleased, but Brenda was curious to know why they hadn't finished the game.

"You still have two more runs waiting," she pointed out. John gently reassured her that it wasn't necessary to finish the inning because the team had already won. But on the next day, when the two teams met for the next game in the series, the Tigers' performance in the first inning supplied that deficiency by generating the two missing runs!

Bob's lifelong devotion to the sport extended to many seasons of university softball competition with graduate students, younger faculty, and technical staff half his age, a pastime that he pursued until he turned 75. In one particular year when Bob's team won their league championship, he sent John a copy of the local newspaper report and John responded with a playful note that read in part: "I have forwarded your contract to the front office for the Tigers. Lately, Lou Whittacker has been suffering back pains. My recommendation is to sign that man, Bob Jahn. As a last place team, we would be proud to win the division title once again."

Our warm relationship with John Fetzer continued until his death in 1991. During those years he continued to provide generous financial support for our research, and invited us to serve on the Board of Trustees of his Fetzer Institute.

4. Snails and Quails

ALTHOUGH OUR proposed research program had already been approved by the university, albeit somewhat reluctantly, by mid-April 1979 Brenda's appointment was still in limbo. The issue appeared to be whether it would be appropriate to appoint a psychologist as a member of the Professional Research Staff to manage an engineering laboratory. The president of the university assembled a distinguished *ad hoc* committee to consider the matter, comprising most of the senior members of the university's administration—the Provost, Dean of the Faculty, Dean of Students, Dean of the Graduate School, and Chairman of the University Research Board. Previously, the Chairman of the Psychology Department had interviewed Brenda and had recommended against her appointment; he felt that since she had indicated that she believed in the possibility of the phenomena to be studied, this disqualified her from conducting "objective" research on the topic.

While we awaited the committee's final determination, Bob was invited to a meeting of the Gardiner Murphy Foundation in Washington D.C., attended by many of the more prominent members of the parapsychology community along with a representative of the McDonnell Foundation, to discuss the future of the field, and Bob invited Brenda to join him. When the meeting convened, the attendees went around the table to introduce themselves and Brenda was the last to speak, following Bob, whose impressive professional credentials were a tough act to follow. Unsure of what she had to offer to the discussion, when her turn came she impulsively described herself as a "low-budget replicator." Given that the two most substantial obstacles facing the field of parapsychology were the needs for financial support and for experimental replication, this response, reflecting her achievements in producing successful replications of the work of others with no financial support, generated a stunned silence in the room as its implications registered with the other attendees.

Following the meeting, we decided to indulge in dinner at an elegant local restaurant as a presumptive precognitive celebration of Brenda's forthcoming, yet still unconfirmed appointment. The restaurant menu featured a special of roast quail, which we both ordered, and Brenda additionally opted for an appetizer of escargot, which did not appeal to Bob, whose tastes ran to more traditional fare. As is typical, the snails were presented in their shells, and Bob observed that it was interesting that all of the shells coiled in the same direction. He asked what she thought was the likelihood of finding one that coiled in the opposite direction. As it happened, not long before Brenda had been assigned this very question on a graduate exam. Frustrated at her inability to answer it, she thought to herself with exasperation, "Who cares which way snails coil? It's not something I'll ever need to know!" Needless to say, she was now dumbfounded by Bob's question!

When they returned to Princeton the next day, Bob left Brenda in the room that temporarily served as the research lab while he attended to his Dean's responsibilities. After about an hour, he reentered the room wearing a very serious expression and handed her an envelope embossed with the university's letterhead. She opened it with a sinking heart, only to discover that her appointment as a research assistant had finally been approved. (One member of the review committee was quoted as saying at that point, "This scares the hell out of me!" but it was unclear whether he was referring to Brenda's appointment, or to the program itself—perhaps both.) Although we had already celebrated this event the evening before, our enthusiasm was reinvigorated and we went off to lunch to make plans for the future.

Brenda would hold that position for the next 28 years.

Tuned In/iStock

5. Christening

ARMED WITH her appointment, Brenda transferred herself to Princeton early in the summer of 1979, accompanied by two children, one dog, two cats, and a plethora of books and personal belongings. Consistent with her stochastic heritage, the move did not go smoothly, but by fall she had acquired a house, enrolled her children in school, and was ready to devote herself to the new enterprise.

At that time the research facilities were situated in two offices on the third floor of the Engineering Quadrangle, but given the ambient noise, busy traffic pattern, and limited space, it became clear that the program would require a more propitious environment. Fortunately, Bob was able to identify an unused storage area in the School basement that he commandeered for the purpose, and arranged for it to be expanded and converted to four small rooms. (A fifth room was added a few years later.) Thanks to a generous gift from a friend of the program, these sterile surroundings soon were transformed into a warm and homey space by installing carpeting, wood paneling, a large orange sectional sofa, and an assortment of pictures and posters, and stuffed animals. A jumble of unmatched metal desks, file cabinets, and bookcases acquired from university surplus, produced an odd contrast to the elegantly appointed reception area.

The PEAR Laboratory

The laboratory did not have a name as yet, and the former storage area had no room number, so a Greek psi (Ψ) was posted on the door, intending to suggest the multiple symbolism of a quantum mechanical wave function, psychology, and psychic phenomena. Over the next several weeks, however, we received several inquiries, accompanied by troubled expressions, asking why there was a devil's pitchfork on the door! The time had obviously arrived for the psi to go and for us to find a more appropriate name for ourselves than "The Psycho Lab," as some of our engineering neighbors referred to it.

Assignment of a label to our novel research enterprise that would inoffensively convey both the essence and substance of its many intellectual, social, and procedural ramifications presented challenges of its own. At one of our working lunches, Bob proposed "Princeton Engineering Anomalies Research," which he contended would economically subsume all manner of pertinent implications. Brenda at first demurred, feeling that the associated acronym was too mundane, but she quickly relented when confronted with the presence on the table of pear-shaped salt and pepper shakers, and when the attending waitress later urged us to try their delicious pear cake for dessert. That constellation of synchronicities left little doubt that henceforth we would be functioning as PEAR.

Clothed with this whimsical nomen, the fledgling program steadily evolved into a scientific enterprise of credible relevance to our liberal arts-based engineering school, and in that context attracted attention from a number of other academic institutions, several of which made overtures to attempt to integrate our approach into their own educational formats. All tongue-clucking notwithstanding, over the ensuing 28 years, the PEAR laboratory continued to grow in local and global respect.

It was early established that in addition to the ongoing experiments and theoretical discussions, our laboratory would welcome all visitors, regardless of the nature or details of their interests, and over the years we entertained a remarkable spectrum of guests. Hardly a week went by without the diversion of lively conversations, fascinating personal stories, and profound philosophical speculations. For many, the PEAR lab represented an intellectual and spiritual refuge where they could share ideas and experiences that were discounted, or even ridiculed, by their professional peers. These exchanges frequently provided us with useful anecdotal insights, such as frequent descriptions of anomalous events being associated with heightened emotional states, or at times of major life transitions. They also contributed to the evolution of an extensive and enduring community of friends and colleagues.

6. Guess Who's Coming To Visit?

THE PEAR LAB may have been small in size, but it was large in heart and was always a lively place. In addition to our regular staff, which over the years ranged between five and nine people, and many groups of visiting school children, there was a constant influx of interns, operators, friends, and visitors who came from a broad spectrum of backgrounds and motivations. Most of these were highly enjoyable and stimulating; others were at least amusing.

One category of visitors comprised representatives of the skeptical community, who came searching for flaws in our protocols and errors in our analyses. At one point, one of these, challenged by our claims of non-local effects, agreed to undertake a series of five remote experiments from approximately 100 miles away. After each set was completed and he had communicated his intentions for each run, we would send him a graph of the results, without comment. The results of the first four sessions displayed highly unlikely effects, albeit in the directions opposite to his intentions. But it must have been evident to him as a scientist that they showed outcomes well beyond chance, to a degree that was also statistically unlikely. He must have become aware of this uncomfortable realization, because his final set had an extremely strong positive trend, sufficient to bring his overall results back precisely to chance. We never heard from him again.

Other guests came eager to demonstrate their putative psychic abilities, most of whom failed to produce any exceptional effects, and to which they occasionally attributed a creative variety of explanations. One claimed that there were negative "geopathic zones" in our laboratory; another maintained that the problem was due to the experiment not being oriented to due east; and one complained that her failure was attributable to the fact that we had coffee brewing in the next room! On more than one occasion we were accused of not creating an appropriately spiritual ambience for the phenomena to manifest. After a few such experiences, we determined that in the future we would avoid involving participants who claimed to have extraordinary psychic gifts. Nevertheless, they were welcome to visit and we were always willing to demonstrate the experiments to them.

Once we entertained a highly respected Chinese qi gong master who claimed to live solely on tomato juice, and who declined to perform a formal experiment that would require our stated minimum of 5,000 trials per intention and take several hours, unless we agreed in advance that he would be first author on any ensuing publication. We explained that this was not an option, but encouraged him to try a demonstration run of 100 trials anyway, which he did. Several years later we learned that he had published a book in which he claimed to have carried out experiments at PEAR with extraordinarily significant results.

Perhaps one of the most amusing such events was a visit from a former physicist, introduced to us by some engineers from a local corporation as a highly enlightened Swami who had heard of our work and wanted to meet us. Prior to his visit, one of his associates contacted us to determine how many people would be attending his seminar. When we pointed out that we would not be sponsoring a formal presentation but merely having him meet informally with members of our staff, he agreed to come anyway. On the day of his visit he marched down the main corridor of the engineering quadrangle in full Swami regalia, arousing considerable curiosity among the administrative personnel whose offices were situated along that hallway, accompanied by an entourage of followers equipped with two cases of recording paraphernalia. He then seated himself in the lab and proceeded to point to the clock on the wall and inform us that we were ignorant of the true nature of time, whereupon he launched into an extensive account of how time was a mere illusion. Bob joined us about halfway through this lecture, indicating after a few minutes by his bemused expression that he regarded this as a waste of time, and he departed shortly thereafter. Precisely one hour after he began, the Swami stopped talking and his attendants began to pack up the recording equipment. As he was leaving, Brenda thanked him for sharing his insights with us and noted that it was unfortunate that we had not had a chance to tell him about our work. She handed him a reprint of one of our publica-tions, which he glanced at and commented, "Ah yes, Jahn and Dunne. I have heard of these people. They are doing interesting work!" She explained that the gentleman who had left was Jahn and that she was Dunne, as she escorted him to the door.

Another individual made it a point to drop by frequently, unannounced, and to insist on engaging us in wearisome conversation. After several months of this, he finally found a job and stopped coming, much to our relief. We almost forgot about him until a considerable time afterwards, as we were breathing a sigh upon the departure of another particularly tedious visitor at 5:00 PM. Brenda observed to Roger Nelson, a member of our staff, "Cheer up, it could have been worse. It could have been ___!" The next morning there was a knock on our door and the former visi-tor walked in, announcing that at around 5:00 PM the previous evening he had decided to spend his day off coming to see us! After this event, we became much more circumspect about making such off-hand remarks.

Baroque Designs Timewarp clock by Julien Hatswel

Over the years we also accommodated a substantial number of TV producers desiring to

The illusion of time

document our findings for the popular media. Unfortunately, very few of them grasped the full nature of our research and most came with preconceived ideas based on what they regarded as the newsworthy assumption that "scientists at Princeton have proven the existence of paranormal phenomena." As a result, all but a few of their programs tended to misrepresent and overstate our results to support their presentations. The fact that the results we had demonstrated were miniscule in scale and only detectable statistically in large databases seemed difficult for most of them to grasp and insufficiently dramatic for their purposes. Notwithstanding, some of these broadcasts attracted considerable attention and resulted in barrages of phone calls and letters from people who wanted to tell us about their own extraordinary experiences, some of which were indeed quite remarkable.

Life at the PEAR lab was never boring.

7. "Tell the Young People"

CONSTRAINTS ASSOCIATED with the university's original approval of the PEAR program precluded any formal academic component or official research involvement with students. Over its 28 years of operation, however, a substantial number of these, as well as many of the thousands of guests who visited the laboratory, were young people who found their way there spontaneously of their own curiosity, yearning for some introduction to this dimension of their being. Some were individuals who had heard of our work and wanted to learn more about it; some wished to share their personal experience and ideas with us; some wanted to sign on as volunteer interns. Many of these became productive short-term members of our staff with whom we maintained ongoing connections as they moved on in their lives.

Early in PEAR's history, one of our interns shared a humorous tape recording containing the repeated entreaty, "Tell the young people!" which became a rallying cry for the PEAR staff whenever such visitors or volunteers appeared, reaffirming the wisdom of the quantum pioneer, Max Planck: "A new scientific truth does not triumph by convincing its opponents and making them see the light, but rather because its opponents eventually die, and a new generation grows up that is familiar with it."[9]

But it was a sequence of younger school groups that proved especially enjoyable and inspiring to us in this context. Specifically, the visits of several fourth grade "gifted and talented" classes from a neighboring elementary school district became a regular and treasured annual event. These 10-year-olds were taking their first science classes, and a visit to a university science laboratory was to be a special part of their program. Their coordinators had approached us somewhat apologetically, having been informed by a number of other laboratories they had contacted for this purpose that their staffs "had no time to waste with a bunch of little kids." In contrast, we had told them that we could not think of anything more important than sharing our work with "a bunch of little kids" and that they were more than welcome to visit.

Over the next decade we were visited by some 1,000 students from the four participating district schools, and each group of about 15 students spent about three hours with us to hear about scientific methodology and basic statistical concepts, try our experiments, and present their own innovative projects. As their teachers became more familiar with

9 Planck, M. (1948). *Wissenschaftliche Selbstbiographie. Mit einem Bildnis und der von Max von Laue gehaltenen Traueransprache.* Johann Ambrosius Barth Verlag (Leipzig 1948), p. 22, as translated in Scientific Autobiography and Other Papers, trans. F. Gaynor (New York, 1949), pp. 33–34 (as cited in T.S. Kuhn, *The Structure of Scientific Revolutions*).

our work, the student projects became more scholarly and sophisticated, and the openness, curiosity, and creativity of these youngsters taught us as much as we taught them. It was especially gratifying to receive an email from one such former student many years later as the laboratory was about to close: she told us that she now was a graduate student at a major university and had been inspired to pursue a career in science by her early visit to PEAR.

Brenda with students

8. Margins of Reality

IN 1985 ONE of our principal sponsors suggested that we consider writing a book that would cogently summarize our studies to date for "an informed, lay audience." Preparation of the book proved to be a major challenge that tested the depth of our molecular bond. Not only was it necessary to cover a large amount of complex technical detail in a readable manner, but it was also necessary to convey the metaphysical dimensions of the material without the book becoming yet another New Age treatise, while maintaining a consistent narrative voice. This was ultimately resolved by having Bob compose the initial drafts of the chapters, which then underwent a number of iterations by both of us, as we wrote, edited, rewrote, and finally came to agreement on content and style.

A major role for Brenda in preparing the book was to identify images that captured the nuances of the written content, as well as suitable sidebar quotations from writers with varied but pertinent perspectives. The multidisciplinary character of the program, and indeed of the topic itself, posed a similar problem of establishing the overall context of the book. After a great deal of discussion and debate, we finally agreed to introduce the main subject matter by an assortment of "vectors" drawn from various traditional areas of human thought and experience that converged on the motivation, definition, and circumscription of this particular research program. The final section of the book revisited these vectors to suggest the implications of the work for each of them.

We had countless disagreements and arguments, and spent endless library hours looking up facts, references, quotations, and pictures. Occasionally, when the work required uninterrupted concentration, we would retreat to more quiet venues to allow ourselves the space to give the book our full attention. Once, on a trip to Florida, Bob returned from a walk on the beach with the comment that he felt the ocean waves were trying to tell him something. This experience inspired the section of the book titled "The Waves of Consciousness." Another involved a stay of several days at Dunwalke, the university's residential guest house in north-central New Jersey—a charming rural mansion in a bucolic setting, attended by a hospitable couple who maintained the facilities and prepared excellent meals for visitors. There we engaged in one of our more volatile quarrels as we attempted to accommodate the many conflicting recommendations offered by the members of our staff who had read the early draft. While we were there, a package of chocolate turtles sent by a friend arrived at the lab, addressed to Brenda. When we returned, she discovered that the package had been opened and its contents devoured by the very people whose comments had triggered our arguments. To put it mildly, she was extremely upset by this and made her displeasure felt.

Difficult as the writing process was, it turned out to be even more of a challenge to identify a publisher willing to consider our manuscript. Fortuitously, on a trip to Dallas, Texas, we had occasion to meet Trammell Crow, a highly respected commercial real estate entrepreneur. Trammell's eclectic interests extended well into many non-traditional topics, including our own area of research, which led to several stimulating conversations. In the course of one of these, he indicated an interest in supporting our work, and inquired if there was anything he could do for us beyond making a financial contribution. When we informed him that we were in the process of writing a book but were having difficulty attracting the interest of any commercial publishers in this project, he offered to introduce us to his personal friend and colleague, William Jovanovich, who presided over Harcourt Brace Jovanovich, one of the nation's leading textbook publishers. This connection led us to a visit to HBJ's Orlando, Florida headquarters, where Bill Jovanovich convened his board of editors to hear our presentation of the proposed book. Predictably, this group politely demurred from endorsing this proposition, after which he promptly adjourned the meeting and invited the two of us to have lunch with him privately at a nearby restaurant. There we were not a little surprised at his opening assertion that if he were to wait until this group of editorial principals came to share our vision, the book would never see daylight, so he would publish it under his own recognizance. He then proceeded to specify quite efficiently some salient financial aspects and editorial suggestions, and assigned his son Peter to preside personally over the details. Then he briskly left us with a jaunty tip of his straw hat.

In subsequent visits to the publisher's San Diego facilities we met with our assigned technical editor, John Radziewicz, who patiently guided us through the three-year editorial process. He graciously made a number of special concessions regarding the design of the book itself, including the use of a cover design that had been hand drawn by Bob (the idea of an author drawing his own cover was quite unusual). Among the added benefits of these visits were the opportunity to have a personal guided tour of the San Diego Sea World facility, to visit to the famed San Diego Zoo, and the use of a private apartment that provided a glorious view of nightly fireworks displays in the distance.

Margins of Reality: The Role of Consciousness in the Physical World was published in an HBJ hardcover edition in 1987, followed by a softcover version the following year. German, Japanese, Russian, and French translations soon followed. In 2009, having obtained the publication rights to the book from the publisher just before HBJ was acquired by Houghton Mifflin, the book was reissued by our own ICRL Press.[10] Notwithstanding the tensions associated with its production, the book remains an excellent example of the strength of our bond, which has been able to withstand our disparities of style to produce a significant and illuminating creation. After more than 27

---

10 R.G. Jahn and B.J. Dunne (2009). *Margins of Reality: The Role of Consciousness in the Physical World.* Princeton, NJ: The ICRL Press. (Originally published in 1987 by Harcourt Brace Jovanovich.)

years, this book continues to sell, primarily via word of mouth since it has never received any substantial media exposure, and is highly regarded as a seminal textbook in the field, widely referenced in this country and abroad.

9. Consciousness and the Source of Reality

BY 2004, PEAR had been in operation for 25 years and several members of our staff, including Bob, were approaching retirement age. Anticipating the closing of our university facilities in the near future, it seemed propitious to undertake a major archiving exercise aimed at documenting and preserving the essential elements of our research program. This included creating backups of all the experimental protocols and data, relevant correspondence, and other important documents, as well as producing a comprehensive video record. Thanks to the skills of Aaron Michels, one of our former interns, a three-part overview of the program was created called "The PEAR Proposition." It included two DVD's that offered four of Bob's lectures to an interdisciplinary undergraduate course in Human/Machine Interactions; a guided tour of the laboratory; interviews with staff members, operators and interns; and sundry other material that might eventually prove helpful to future scholars. It also included a CD recording of a conversation between the two of us wherein we discussed the motivations, goals, and dynamics that had guided the program. A comprehensive article with the same title was published in the *Journal of Scientific Exploration* in 2005.[11]

Other archiving tasks still remained, including the preparation of a sequel to *Margins of Reality*, which had documented only the first eight years of the program. In the subsequent 20 years we had acquired a substantial amount of additional data, developed some new conceptual models, and recognized that there were several contextual vectors that had not been included in the first book but which had become more relevant over the ensuing years. More importantly, as we probed further into the nature of these anomalous phenomena, we had become increasingly persuaded that they were rooted in, and indicative of, a much more fundamental, profound, and ubiquitous metaphysical dynamic whose ultimate comprehension held far richer potential than the explicit phenomenological curiosities with which we had begun. Indeed, in prompting epistemological penetration beyond the superficial "margins" of reality, these studies had led us into contemplation of their essential "Source," and provided a glimmer of a vast, poorly charted domain for future study that had profound implications for understanding the nature of consciousness itself.

Thus, we embarked on another creative editorial collaboration that ultimately led to *Consciousness and the Source of Reality: The PEAR Odyssey*, which

11 Jahn, R.G. and B.J. Dunne (2005). "The PEAR Proposition." *Journal of Scientific Exploration*, 19 (2) pp. 195–246.

was published by our own ICRL Press in 2011.[12] Following the general format of *Margins of Reality*, this book introduced a number of new vectors: biology, medicine, information technology, human creativity, and the spiritual roots of science. These were followed by a detailed reportage of all the empirical studies performed in the PEAR laboratory, descriptions of several theoretical efforts, and concluded with considerations of the nature of information, uncertainty, and the limitations of contemporary science. This time our collaboration was much less discordant and the preparation went much more smoothly. Once again we found that the essential complementarity of our respective interests and insights was able to foster a work that neither of us could have produced independently. Consistent with the molecular nature of our relationship, it was difficult to determine from the end product just who had contributed what to the ultimate product.

---

12 R.G. Jahn and B.J. Dunne (2011). *Consciousness and the Source of Reality: The PEAR Odyssey.* Princeton, NJ: The ICRL Press.

10. Quirks of the Quantum Mind

GIVEN BRENDA'S lack of any formal background in theoretical physics, she often found herself somewhat intimidated by the tendency of many physicists, when asked to explain certain basic concepts, to respond with a statement to the effect of "Let me give you a trivial example of hydrogen." One day she challenged one of her physicist colleagues to explain the difference between "trivial" and "non-trivial," and he pointed out that "trivial" simply meant that the concept could be solved in principle, whereas "non-trivial" indicated that it was impossible. She was somewhat shaken to realize that physics was unable to account mathematically for anything more complicated than hydrogen. This emphasized for her the degree to which most physical principles were simply constructions of consciousness.

In an effort to help Brenda understand the principles underlying quantum mechanics, Bob invited her to sit in on one of his graduate classes, assuring her that she need not worry about the mathematics but simply try to get a grasp of the fundamental concepts. When one of the students in the class asked her why a psychologist would want to learn about quantum mechanics, she responded with mock surprise, "Quantum mechanics? I thought this was about the psychology of atomic particles!"

The two of us shared many stimulating dialogues about the metaphorical relevance of quantum mechanical concepts for addressing the nature of human consciousness, a conversation that continued throughout the many years of our relationship. Exploring these from our respective complementary perspectives, we came to realize that quantum mechanics, indeed any model of physical reality, is not so much a description of the physical world per se, but a product and reflection of the human mind as it attempts to represent both its tangible observations and its subjective experiences of it. Together we extended these ideas to develop an extensive spectrum of quantum mechanical concepts and principles, and to deploy these in a more comprehensive metaphorical model of the processes of consciousness. This exercise developed into a language and viewpoint we found usefully applicable to many physical and psychological phenomena. For example, we might comment that the "angular momentum" of a given situation had a "positive spin," or that our "eigenfunctions" were especially "resonant." Our identification of Bob as the "particle" and Brenda the "wave" components of our molecule, proved very apt, and by reminding us of the complementary nature of these qualities it encouraged us to be respectful of each other's perspectives.

Curious to know how the early founders of quantum mechanics might have regarded such an idea, we examined many of their philosophical publications.

Although these books were easily available in the university physics library, it soon became evident that very few of them had been consulted over the years. While most contemporary academic physicists extensively employ the mathematical and analytical tools of these pioneers, they seem to be less familiar with the philosophical reflections that underlie them. This material proved to be fascinating reading and confirmed that many of these pioneers had early recognized that consciousness was an important component of physical reality, and thus of its theoretical models.

As in the conceptualization of the complementarity of wave and particle, each of which is correct in its own observational frame of reference, most of these pioneering scientists had acknowledged that the constructs of physics and psychology were not mutually exclusive, but were complementary representations that do not exclude one another. As expressed by Wolfgang Pauli's observation that: "…the idea of complementarity in modern physics has demonstrated to us, in a new kind of synthesis, that the contradiction in the application of old contrasting conception (such as particle and wave) is only apparent; on the other hand, the employability of old alchemical ideas in the psychology of Jung points to a deeper unity of psychical and physical occurrences."[13]

---

13 W. Pauli (1955). "The Influence of Archetypal Ideas on the Scientific Theories of Kepler," in C.G. Jung and W. Pauli. *Interpretation of Nature and the Psyche.* (Tr. P. Silz). NY: Pantheon Books (Bollingen Series LI), p.207.

Nearly all of them had interests in various forms of Eastern mysticism and/or Western philosophy, and they had discussed these dimensions at length.[14] Several of them had even written extensively on the implications of how the physical principles they had formulated reflected the central role of consciousness in comprehending the fundamental nature of reality.

When we had first developed our own model in the form of a technical report, we had included an extensive Appendix containing numerous quotations from the many of them, including Planck, Bohr, Heisenberg, Pauli, de Broglie, Schrödinger, Einstein, Wigner, and several others.[15] In 1985 we submitted our model as an article (without Appendix) to the journal *Foundations of Physics* for publication and, as is standard with academic journals, the editor initiated the usual peer review process. What was not standard was that this article ultimately was so reviewed by 17 referees (the typical procedure is to send it to two or three at most), including several Nobel laureates. There were strong reactions, both positive and negative, and we finally suggested to the editor, somewhat facetiously, that he count the responses and go with the majority. With nine "ayes" and eight "nays," the article was finally published in August 1986, in a slightly

---

14 W. Heisenberg (1971). *Physics and Beyond: Encounters and Conversations* (Tr. A.J. Pomerans). NY: Harper and Row

15 R.G. Jahn and B.J. Dunne (1983). "On the Quantum Mechanics of Consciousness, With Application to Anomalous Phenomena." *PEAR Technical Report* #83005.

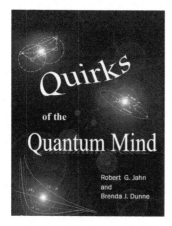

abbreviated form, with the title "On the Quantum Mechanics of Consciousness, With Application to Anomalous Phenomena."[16]

Although this model was summarized in both *Margins* and *Source*, much of the feedback we received suggested that readers had misunderstood our main point, assuming that we were positing a quantum mechanical explanation for consciousness, rather than proposing that quantum mechanics was itself an indicative product of human consciousness. In 2011, therefore, we prepared a clarified and extended version of the original article, together with its extensive Appendix, which we published in book form under the name *Quirks of the Quantum Mind*.[17]

---

16 R.G. Jahn and B.J. Dunne (1986). "On the Quantum Mechanics of Consciousness, With Application to Anomalous Phenomena." *Foundations of Physics*, 16, No.8, pp.721-772.

17 R.G. Jahn and B.J. Dunne (2011). *Quirks of the Quantum Mind*. Princeton, NJ: The ICRL Press.

11. The "Meds"

PERHAPS BECAUSE of their higher analytical dispositions and trainings, perhaps because of their more frequent encounters with paradoxical physical events in their own career practices, those members of the medical profession who have interacted with us in the course of our professional activities have tended to be respectful of our empirical results. This has facilitated many productive conversations with them bearing on the anomalous aspects of our work and theirs.

What immediately comes to mind are the immense accomplishments of Dr. Larry Dossey, who set aside his career as a senior medical practitioner at a major Dallas hospital in favor of one devoted to speaking and writing about various medical anomalies bristling with health-related ramifications. He has authored many books about alternative health care and spirituality that have had far-ranging impact on the practice of modern medicine and on the public at large.

At one point, Larry invited us to give a presentation about our work to a large group of medical people. We were happy to do so, although we weren't sure whether the medical community would really be interested in our anomalies research. But to our amazement, following our talk the audience gave us a standing ovation! Later, when we asked why they had responded so enthusiastically, Larry pointed out that these people witnessed medical anomalies quite frequently in the course of their work but felt they couldn't talk about these openly because there was no "scientific evidence" to support them. Apparently, we had just presented them with such evidence. As Larry states on his website: "Almost all physicians possess a lavish list of strange happenings unexplainable by normal science. A tally of these events would demonstrate, I am convinced, that medical science not only has not had the last word, it has hardly had the first word on how the world works, especially when the mind is involved."[18]

Larry Dossey is not alone. In our local community, Bob has been treated and healed by Dr. William Haynes, a cardiologist who has maintained an ongoing interest in our work. He routinely prays with and for his patients, and has published two books describing his personal philosophy and professional experiences.[19,20] If the exponential proliferation of professional articles and popular literature on this and related topics is any indication, the role of consciousness in health and healing is becoming increasingly acknowledged by the medical community as a subject to be taken seriously.

---

18 http://www.larrydosseymd.com

19 William F. Haynes, Jr. (1990). *A Physician's Witness to the Power of Shared Prayer.* Lincoln, NE: iUniverse.com, Inc.

20 William F. Haynes, Jr. and Geffrey B. Kelley (2006). *Is There a God in Health Care?: Toward a New Spirituality of Medicine.* Binghamton, NY: The Haworth Press.

In the course of our investigations we have had extensive interactions with a variety of medical practitioners, psychiatrists, and psychologists who were seeking some understanding of the strange events that their patients, and they themselves, had encountered in their clinical practices. Not surprisingly, they had inevitably conceded the relevance of our studies to their own experiences. To these folks, such medical anomalies as placebo effects, allergic reactions, multiple personality syndromes, and spontaneous healings have not been idle curiosities. Rather, they have been recognized as persuasive evidence of our currently incomplete comprehension of the nature of living systems, and cannot simply be swept under the medical rug.

A traditional responsibility of the Dean of Engineering at Princeton has been to serve as one of two senior faculty delegates to a consortium of academic institutions who have bonded together to avail themselves of major physical and astronomical research facilities that would be financially inaccessible individually but that lend themselves to cooperative utilization and management for major projects in the high sciences and engineering.[21] Bob served in this role for several years, including being elected to chair this distinguished group. Although the majority of its members were drawn from the faculties of the "exact" sciences of their respective institutions, some of them, like Dr. William Sweet of the Harvard Medical School, represented such tangential disciplines as

---

21 The Associated Universities, Inc.

the medical sciences. A man of piercing penetration of judgment, no-nonsense style, and dry humor, Bill endeared himself to Bob—as Bob to him—in his hard-nosed attention to the affairs of the consortium. Given Bill's outstanding professional work in pain management, consciousness research became a natural topic for their private discussions. As a consequence, he and his wife invited the two of us to visit them in Boston for the purpose of having us speak about our work to two separate groups. Brenda spoke to an informal evening audience of some of their personal acquaintances with interests in the metaphysical dimensions of the topic, and the following day Bob gave a more technical version to an assembly of Bill's professional colleagues in the renowned "Ether Dome" at their medical school.

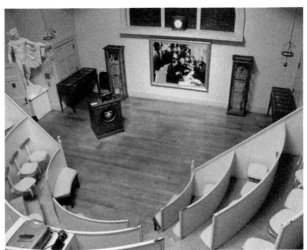

The Ether Dome, Harvard Medical School

Elliot Joel Bernstein

Each of these talks offered a different perspective and style, yet the audience receptions to both were genial and enthusiastic and generated a number of insightful questions, as well as several fascinating reports of personal and professional encounters with anomalies. This genuine interest in our research continued to characterize our interactions with the medical communities throughout the course of our program.

In 2007, as a parting tribute to the PEAR program, *EXPLORE: The Journal of Science and Healing* published a special 150-page issue dedicated entirely to an anthology of that portion of the PEAR work that bore relevance to the field of complementary and integrative medicine.[22] The

issue was titled "The Pertinence of the Princeton Engineering Anomalies Research (PEAR) Laboratory to the Pursuit of Global Health," and comprised reprints of 10 articles and 12 abstracts we had published over our 28-year history. Wayne Jonas and Harald Walach,

---

22 "The Pertinence of the Princeton Engineering Anomalies Research (PEAR) Laboratory to the Pursuit of Global Health," *EXPLORE: The Journal of Science and Healing.* May 2007, Volume 3, Issue 3, pp.191–346.

Director and European Director of the Samueli Institute, respectively, along with Larry Dossey, the executive editor of *EXPLORE*, all of whom are acknowledged leaders and spokespersons for this field, prefaced the issue with their own personal endorsements.

12. The "Feds"

AS WORD OF THE PEAR program spread, we were approached by many individuals and institutions that expressed interest in its activities and findings. From the outset, we had determined that in order to be free to publish any of our findings in the open literature, we would accept no government funding or undertake any classified research. Nonetheless, we entertained visits from representatives of various government offices and agencies (some of whom identified themselves as such, and some who did not). It quickly became evident that these organizations were quite interested in what we were doing and were watching us closely, and several of these invited us to give presentations to their staffs.

One such request was for Brenda to speak on the topic of remote perception at the National Security Agency. This was early in the program and Brenda, who had never has occasion to give a talk to that class of audience, was understandably quite nervous. To ease the tension, she began by projecting a cartoon image of a dragon at a podium, addressing a room of armored knights and announcing that, "while there are still profound differences between us, I think the very fact of my presence here today is a major breakthrough." The laughter this evoked relaxed the atmosphere in the room noticeably, and her stage fright dissipated.

We also spoke at several NASA locations and gave presentations at the Naval Research Laboratory and elsewhere, including a meeting of ranking officers of the U.S. Marine Corps who were exploring possible future technologies. On most of these occasions our talks were received politely, albeit cautiously. Yet, as in virtually all of our public lectures, after most of the room had cleared a few attendees would remain behind and approach us rather self-consciously. After first looking over their shoulders to assure privacy, they would confide that they had had unusual personal experiences themselves, which they would like to share with us.

Another noteworthy government interaction involved a private meeting with two members of the White House staff who wanted to know if remote perception could enable an individual to distinguish a silo containing a live missile from an array of decoys. (This was during the period when our government was contemplating establishment of a broadly based MX missile program.) We opined that this would probably pose no serious problem and noted that viewers could probably also determine the future site of such a live missile if it were to be moved. Shortly after this interaction, we decided to set up our own small pilot study that involved randomly placing an array of opaque film cans on a large map of a southwestern state. Each can contained a ¾" black marble, along with one can that contained a red marble of the same size and material. The cans would be placed at random locations on the map, and a percipient would attempt to identify which one contained the red ball. We playfully named this project "Missile, Missile, Who's Got the Missile," but this experiment never proceeded beyond the planning stage before the government's MX program was aborted.

There were numerous other government-related seminars and conferences, as well as personal interactions with a few Senators, Congressmen, and other representatives of various federal agencies, several of whom were attempting to promote basic research in this area and whom we came to regard as personal friends. One of these was a ranking U. S. naval officer who maintained a keen personal interest in the topic and had been assiduously graceful in arranging our invitations to several of these venues, and for many years we deeply benefited from our relationship with him. For example, in one instance when we were experiencing some difficult issues with our university, he arranged for a senior member of the research and development directorate of the Department of Defense to visit the university president personally to express an official sense of the importance and relevance of our research to the national R & D efforts. While we were not present at that interview, it did seem to relax some of the prevailing academic distaste for our work.

On other occasions, when Bob had to travel to Washington for meetings associated with his aerospace work, Brenda would accompany him, and we often took those opportunities to meet with our Navy friend. At some point during these meetings, he would somewhat awkwardly ask Brenda to leave the room.

(These private discussions typically dealt with classified matters, and while Bob held a high clearance level, Brenda had none.) She found these situations amusing and pointed out that it seemed a bit silly to restrict her access to covert information about such topics, given the nature of the phenomena! On one such occasion Bob apologized for her, explaining that "she talks funny, but you'd better listen."

As time passed, government attention to this topic seemed to wane, and these interactions became less frequent, although we assumed that the Feds continued to maintain some cognizance of our activities. We suspected that our attempts to explain what we believe underlies the anomalous phenomena to people committed to a deterministic paradigm may have been too problematic for them to digest comfortably.

Alexandra Huyghe

13. Professional Societies and Skeptics

ALTHOUGH THE two of us had first met at a meeting of the Parapsychological Association (PA) and had published one of our early research articles in its journal, after attending a few more of its annual meetings it became evident that our scholarly path diverged significantly from that of parapsychology. This was primarily because most of the researchers in that field had professional backgrounds in experimental psychology and tended to deploy methods and protocols that emphasized the characteristics of individual "subjects" and their cognitive strategies, while the PEAR program, with its engineering perspective, focused on the behavior of random physical systems and processes. As a result, the hypotheses, experimental designs, and data processing techniques we employed differed substantially from those of parapsychological research.

Another factor distinguishing our work from that of parapsychology was that field's tendency to emphasize the "paranormal" nature of the phenomena it addressed, while we preferred the term "anomalous" since we regarded the phenomena as ultimately normal, albeit currently inexplicable. Notwithstanding Bob's having served for many years on the Boards of the Rhine Institute and the American Society for Psychical Research, we eventually stopped participating in PA meetings and contributing to its journals. This caused some members of the parapsychological community to regard us as elitist, and our interactions with them became rather cool. When the *Journal of Parapsychology* published an article sharply critical of our remote perception research that reached an *ad hominem* attack, we decided it was time to part ways.

This difference in emphasis also was reflected in the nature of our interactions with the skeptical community. Whereas many parapsychologists devoted extensive effort to arguing, and sometimes even to collaborating with professional "debunkers," we preferred to minimize these interactions, responding to their criticisms only when they were directed specifically to our protocols, analytical methods, or empirical results, and could be directly rebutted with scientific authority. We were nevertheless a frequent target of these critics, who attempted to denigrate, or even to misrepresent our work.

On one occasion, our university extended an invitation to James Randi, a professional magician and one of the most outspoken detractors of any topic that he and his skeptical associates regarded as "pseudoscience," to give a distinguished lecture on the Princeton campus. On this occasion, Randi visited our laboratory and attempted, without success, to detect flaws in our experiments. Nonetheless, we maintained a cordial interaction with him at that time and displayed an affable group demeanor at the reception and dinner held in his honor. In its coverage of his talk later that evening, *The Daily Princetonian* quoted him as saying that he was "impressed with the psychic research being done at Princeton. There's very interesting stuff going on here. I have more hope in it than I've had for anything in a long time."[23] (Pleased as we were by the compliment, we were understandably amused by the unusual number of anomalous malfunctions of the various technical devices he used in his presentation.)

A more productive opportunity for collegial interaction arose from Bob's 1977 visit with Peter Sturrock, a professor of astrophysics at Stanford University. Peter was interested in establishing a multi-disciplinary professional society that would provide a forum for scholars interested in a fairly wide range of topics that were typically rejected or ignored by mainstream science. Together with a number of other senior researchers, in 1982 the group established the Society for Scientific Exploration (SSE). Since its inception, both of us have been actively involved in SSE, serving on its Council and Executive Committee, publishing numerous articles in its peer-reviewed *Journal of Scientific Exploration*, helping to organize many of its annual meetings, and coordinating its educational activities. Bob served as Vice President,

---

23 A. Schneider (1984). "Amazing Randi amuses crows with tales of alleged psychics.) *The Daily Princetonian*, February 15, 1984.

and Brenda as Education Officer for many years, and Bob presented the first SSE "Founders Lecture" at its 34th Annual Meeting in 2015.

Bob giving SSE Founder's Address

14. A Night at the Opera

AS TIME WENT on and our relationship deepened, we came to share more and more of our personal interests with each other. One of Bob's long-time preoccupations had been with opera, especially the works of Richard Wagner. At the time the only thing Brenda knew about Wagner was through her studies of the philosopher Friedrich Nietzsche and his relationship with Wagner. But when Bob introduced her to the story of *Der Ring Des Niebelungen*, with its deep mythic underpinnings and sophisticated musical structure, she was immediately fascinated. Not long after that, they had an opportunity to attend excellent performances of the entire four-part Ring cycle at the Metropolitan Opera, and Brenda instantly became an opera fan.

From that point on, we made it a point to attend several operas each year, a diversion we both enjoyed immensely, and which provided yet another shared dimension to our personal bond. In later years, we often brought a small portable experimental device with us to these events and found some strikingly anomalous effects, especially when the performances were particularly engaging.

Although our interests in the genre extended well beyond the works of Wagner, it was one of Wagner's other operas, *Die Meistersinger von Nürnberg*, his only comic work, that became our personal favorite. The story is set in 16th-century Nuremberg, where Walther, a young knight, woos Eva, a beautiful young woman whose father is a member of a Guild of Mastersingers and has promised to give his daughter in marriage to the winner of the annual songfest competition. In order to pursue his quest for Eva's hand, therefore, Walther must become a Mastersinger. But the Guild has very specific rules and strict regulations on what should and should not be in a song, and Walther is not familiar with them. So when he auditions for the Guild he breaks most of the established rules and is ridiculed by the Mastersingers, and his membership application is rejected.

The main character in the story, however, is Hans Sachs, a respected member of the Guild, who is based on a real-life shoemaker-poet who also wrote music and plays. He sees some promise in the young knight's new kind of song and decides to help him. With some tutoring by Sachs, Walther manages to recast his song in a way that incorporates the old traditions, but still maintains its innovative style. Ultimately, Walther wins the song contest and the girl.

Many aspects of the plot of this opera bore strong similarities to our own academic situation.

Music from *Die Meistersinger*

Nationalarchiv der Richard-Wagner-Stiftung

Like Walther, we had attempted to introduce a new concept to the professional "Guild," only to meet with scorn and rejection. One respected member of that community (namely the Dean of Engineering and an esteemed rocket scientist), however, was visionary enough to recognize the intellectual potential of this new "song," and to represent it in a manner consistent with the traditional scholarly method. At the same time, the newcomer with the wild ideas had to learn the prevailing academic and scientific rule systems and accommodate to these. Not least of all, it was impossible not to resonate with the plot's underlying humor. Whether the PEAR opus will ultimately win the "prize" of scientific respectability is still undetermined, and we currently await the final act of our version of this masterpiece.

15. Hither and Yon

OVER THE YEARS, professional meetings and conferences provided numerous occasions for us to travel together to a wide range of domestic and international venues. Not only were these convocations interesting and informative in themselves, they also afforded opportunities for us to make new friends and colleagues, and to develop a productive scholarly network. Most importantly, these trips enabled us to engage in intense personal discussions, to pursue promising theoretical possibilities, and to acquire stimulating insights that reinforced our personal and professional bond, and several of them encompassed experiences of a profound metaphysical nature. On a few occasions these trips also provided opportunities to carry out remote perception experiments with percipients back in the U.S.

In some of these earliest trips, Brenda accompanied Bob to several meetings of some his professional organizations, including the annual gatherings of engineering deans and their wives. Although she enjoyed these opportunities to meet some of his colleagues, she often felt somewhat out of place in this company. On one such occasion in San Antonio, Texas, we joined several of the meeting participants on a boat trip on the San Antonio River. Looking up at the beautiful canopy of overhanging trees, Brenda exclaimed appreciatively, "Wow! Look at those trees!" and then became aware of several of the women in the boat staring at her with evident disapproval. Mustering a pleasant smile, she managed to recover her poise, realizing that she would need to learn how to rein in her enthusiasm if she wanted to be taken seriously by such an academic group.

On our trips together we discovered that we both preferred natural landscapes and ancient ruins to the standard urban attractions of museums, cathedrals, and shopping districts, and our travels enabled us to explore many of these more inspiring locales. For example, a meeting in Delphi, Greece, provided us with an opportunity to see many ancient sites and to take an Aegean cruise that stopped at several magnificent islands. At both Eleusis and Patmos we were personally greeted by cats, who appeared to be expecting us and undertook to lead us around the sites. Both of these originally were sacred precincts dedicated to the Mother Goddess, and on the plane returning home Brenda took out a book by C. G. Jung & C. Kerényi[24] that she had brought along, and opened it at random to a page that stated that the Mother Goddess frequently appeared to Her devotees in animal form, often that of a cat!

On a memorable trip to the U.K. we rented a car and spent a week driving on the "wrong" side of the road as we toured England and Scotland. We visited Sherwood Forest, where we encountered the

---

24 C. G. Jung & C. Kerényi (1969). *Essays on a Science of Mythology: The Myth of the Divine Child and the Mysteries of Eleusis*. Princeton University Press.

Ruins of Eleusis

venerable Major Oak, said to have afforded a hiding place for Robin Hood and his merry band; crossed Brigadoon; scanned Loch Ness unsuccessfully for a glimpse of Nessie; visited an assortment of ancient castles, cairns, and little-known stone circles; heard a kilted piper play "Scotland the Brave" at dusk from a turret of Edinburgh Castle; and explored the wonders of Stonehenge, Avebury, and Glastonbury.

The Major Oak, Sherwood Forest

On our way to Heathrow Airport for the return trip home, we lost our way and seemed to be travelling in repeated circles. Brenda retains a fond memory of Bob, who seldom lost his unflappable cool, uncharacteristically shouting angrily, "I don't want Woking; I want Dorking!" as he attempted to navigate a map (an expression of frustration he has occasionally been heard to repeat in moments of stress.) Years later we enjoyed two additional trips to the U.K., one to explore the acoustical properties of several prehistoric ceremonial sites (cf. Chapter 16), and another to visit several islands of the Inner Hebrides, including Tiree and Iona.

In 1987, Bob was awarded an honorary doctorate from Andhra University, which afforded us an opportunity to spend a few days in India, where we visited the Taj Mahal, several intricately carved old temples, and many other impressive sites. While there, Brenda enjoyed wearing a sari and was frequently taken for a native being escorted by a strange foreign-looking man in a red baseball cap. Many years later, following an SSE meeting in Norway, our friend and colleague Erling Strand took us on a personal tour of its impressive fjords and other high country wonders, and we enjoyed similar escorted excursions while in Germany, Italy, Spain, Denmark, and the Netherlands.

We have been fortunate as well to have visited many of the splendid National Parks and scenic vistas in various parts of the U.S. One of our favorite destinations, to which we have returned on several

Norwegian fjord

occasions, is the Four Corners area of the southwest with its Grand Staircase of Zion, Bryce Canyon, and Grand Canyon National Parks and the spectacular surrounding areas. On our first trip there, we had to decline an invitation to attend a traditional Hopi rain dance in order to catch our flight home the next day. Driving toward the airport, we could see the Hopi Mesas in the distance, perhaps 40 or 50 miles to the north, and noted that while the surrounding sky was clear, there was a pronounced cluster of dark storm clouds hovering directly above the mesas where it was raining heavily. On a subsequent visit to the area we recalled that earlier experience and began to speculate on how one might invoke such an event. As we discussed this, the sky above us grew dark and we found ourselves in the midst of a substantial downpour, although it was not the rainy season. Curiously, each time we arrived at one of our intended destinations the rain would stop and permit us to visit for an hour or two before it began again. This happened frequently enough that we began to refer to "Storm" as if it was a sentient being, deliberately following us. At one point, when we were driving through a beautiful area known as the Vermillion Cliffs, the rain again suddenly subsided and we found ourselves near a Native American gift shop tucked into the cliffs. We went inside, where Brenda purchased a necklace with an inlaid image of a Kachina. When she inquired of the proprietor which Kachina it represented, the woman replied. "Oh, that's the Storm Kachina." As soon as we got back in the car, with Brenda wearing the necklace, Storm set in again with great intensity, producing an impressive display of sheet lightning in the distance. It abated as soon as we reached Monument Valley, the final destination of our trip, but shortly after we drove into the park our car became mired in a deep mud puddle and we were unable to proceed. Eventually, a Park Ranger arrived in a truck and attempted to tow us out, but his tow chain broke and he had to call for reinforcement. By the time we were finally extracted several hours later, we and our car were covered with red mud and we could easily imagine Storm being amused at our predicament.

On a visit to Chaco Canyon with our colleagues Paul and Charla Devereux, we experienced another sequence of anomalous events. Paul, an author and specialist in the study of ancient landscapes, was looking for evidence of a stairway in the canyon cliff that could have been associated with one of the mysterious straight tracks in the regional landscape. While the others were walking around staring at the walls of the cliff for hints of the stairway location, Brenda wandered off a short way to approach three ravens perched on top of the cliff who seemed to be calling her. When she reached the spot where they were sitting, two of them flew off and the third cawed loudly, revealing the missing staircase directly in front of her! Later during that same expedition, we parked our car and walked toward an area of ruins, taking many pictures. Although the camera had indicated that there should have been only four shots left on the current roll, the film continued to advance. "This reminds me of one of those conical-shaped things that constantly refills itself," Brenda commented to Charla, unable to recall

Bill Stine

Ancient steps in Chaco Canyon

the name of the object she had in mind. But a few minutes later they encountered a Park Ranger wearing a nametag that read "Mr. Cornucopia."

Anomalous occurrences like these were almost common on many of our travels. Unlike the phenomena we were studying in the laboratory, they were inevitably spontaneous and were almost always associated with our mutual sense of wonder at the grandeur of the natural world, but also with the beauty and magnitude of our interpersonal bond, which became especially strong when we were in these awe-inspiring sites.

16. Archaeoacoustics

SOMETIME IN the early 1990s, on another trip to the American southwest with Paul and Charla Devereux, we stopped in Cortez, New Mexico, to visit the reconstructed ruins of the ancient Great Kiva, an impressive large circular structure with a ring of

The Great Kiva, Cortez, New Mexico

outer rooms, or niches. While there, Bob observed that the site seemed to produce exceptional acoustical resonances and he asked whether anyone had ever studied its acoustical properties. Paul responded that he knew of no such studies but indicated that if we were interested in the acoustics of ancient structures we should visit some of the prehistoric sites in Great Britain. Intrigued by this prospect, we proceeded to make plans to visit some of these to carry out some rudimentary measurements.

We planned a modest itinerary, assembled a rudimentary equipment package, and a designed a simple protocol that would constitute a pilot experiment. Over a 10-day period during July 1994, along with Paul and Charla we visited six ancient sites in the U.K.—five of which dated to 3,500 B.C. or earlier—and measured the frequencies of the acoustic standing waves supported by each of them.

Notwithstanding the substantial irregularities in the external and internal shapes and surfaces of these various structures and their varying degrees of deterioration, we were surprised to note that the resonant frequencies in all of them were well-defined, lying within the narrow interval between 95 and 120 Hz, well within the range of the adult male voice. Even more surprising was the observation that the extensive rock art at some of the locations displayed striking similarities to the standing wave patterns that characterized these chambers. Traditionally, these etchings of concentric circles, ellipses, or spirals have been regarded as astronomical indicators, and indeed

Paul Devereux

Bob measuring sound patterns

most of these sites have been shown to be correlated with prominent solar or lunar events, but they had not previously been associated with the acoustical properties of the structures themselves. A number of them featured sinusoidal or zig-zag patterns that corresponded to the alternating nodes and antinodes of their specific chambers. Those at Newgrange, in particular, had precisely the same number of nodes and antinodes as the resonant standing wave patterns we mapped along the long entry passage, and one of the

kerbstones outside the Newgrange structure displayed
a pattern that in number and spacing were accurate
representations of these nodal patterns. In a few cases,
it even appeared that various interior standing stones
had been positioned as acoustical baffles to enhance
the resonances.

Kerbstone markings at Newgrange

These preliminary and relatively unpreten-
tious observations left little doubt that the structures
themselves had been built with a deliberate intent
to produce specific acoustical resonances, and that
their builders had a sophisticated understanding of
the nature of sound. How the structures were utilized,
however, and for what ceremonial purposes, is still
unknown. Yet, standing outside the structures while
these reverberations were taking place, one could easily feel the vibrations in one's body, and it seemed

that they even might be affecting the neighboring environment as well. For example, the cows in a near-by field appeared to be lowing at essentially the same frequency!

Our report of these studies was published in a prominent engineering acoustics journal,[25] as well as in a respected archaeology periodical.[26] Unlike the controversial reactions we encountered in response to our PEAR anomalies research, this modest investigation generated an enthusiastic reaction that has prompted many scholars to undertake similar studies at other ancient ceremonial sites, and has stimulated the establishment of a dedicated refereed journal[27] and the formation of a professional society that now holds annual meetings. In 2001 the BBC produced a televised documentary on the topic for which Paul prepared a companion book.[28] And in 2012, the prestigious American Association for the Advancement

---

25 Jahn, R.G., P. Devereux and M. Ibison (1996). "Acoustical Resonances of Assorted Ancient Structures." *Journal of the Acoustical Society of America*, 99 (2) pp. 649–658.

26 Devereux, P. and R.G. Jahn (1996). "Preliminary Investigation and Cognitive Considerations of the Acoustical Resonances of Selected Archaeological Sites." *Antiquity*, 70 pp. 665–666

27 *Time & Mind: The Journal of Archaeology, Consciousness and Culture* (Berg Publishers)

28 Paul Devereux and Tony Richardson, *Stone Age Soundtracks: The Acoustic Archaeology of Ancient Sites*, Vega, 2001.

of Science (AAAS) sponsored a presentation on the topic at its Annual Meeting.[29]

As a sequel to our investigation, one of our colleagues, Dr. Ian Cook of UCLA's School of Medicine, along with some of his associates, conducted a pilot study to examine the effects of frequencies in the range of approximately 110Hz on brain activity.[30] (This is the frequency of the musical note "A" played two octaves below the 440 Hz "tuning A" routinely used by orchestras since the adoption of the tempered musical scale in the 1700s.) The central finding in this pilot study was that listening to tones at 110 Hz was associated with patterns of regional brain activity that differed from those observed when listening to tones at neighboring frequencies. These differences were statistically significant in areas of the brain that have been implicated in the cognitive processing of spoken language, which would be consistent with deactivation of language centers in the brain under that condition to allow internal mental processes, such as daydreaming, to become more prominent. They suggest that these structures may have played a role in generating altered-state ritual-driven experiences.

The authors of that article noted that, "While this may be simply a curious coincidence, it is also

29 https://aaas.confex.com/aaas/2012/webprogram/Session4972.html

30 Cook, I.A., S.K. Pajot, and A.F. Leuchter (2008). "Ancient Architectural Acoustic Resonance Patterns and Regional Brain Activity." *Time and Mind: The Journal of Archaeology Consciousness and Culture*, 1 (1) pp. 95–104.

possible that the development of the Western musical scale could reflect some intrinsic properties of the human brain and mind, and the acoustic properties of the Neolithic structures which began this inquiry may have been selected to couple into these brain mechanisms, even if the designers of these structures had only an empirical understanding of the phenomenon."

While others have continued to conduct research on this topic, we have not pursued it further. Nevertheless, we were gratified to have at least this one aspect of our work regarded as acceptable by the scholarly community!

17. Animal Magic

OUR SHARED LOVE of nature and animals has been a deeply meaningful component of our inter-personal molecule. Bob has had several wonderful dogs and two cats over the years, as well as a par-rot, and Brenda has had a dog and a number of cats, and we have shared our affection for all of them. We have been fond of taking frequent walks in the various woodlands and parks in the Princeton environs and elsewhere, where we have enjoyed some of our most productive conversations, and where we have had recurring encounters with a wide variety of wild animals and birds with whom we have interacted, sometimes profoundly. For several years, before the local authorities banned the activity, we took delight in feeding the Canada geese that inhabited a small lake in a local park. Over time, one pair of these birds came to recognize us and would swim to greet us, and even allow us to hold their goslings—a very unusual behavior for these wild creatures. On one such occa-sion, when the adults detected the approach of a dog, they issued a cautionary cry to their babies to get back to safety in the water. The young geese obeyed, but the parents looked at us intently and repeated their alert, apparently informing us of the approaching danger and expressing concern for our safety.

On one of Brenda's early visits to Princeton, before the program had been formally established,

Brenda with geese

Bob took her to visit the woods associated with the
renowned Institute of Advanced Studies. We stopped
in a peaceful nearby field where we engaged in a deeply
meaningful conversation, during the course of which
Bob described how he had become drawn to the study
of anomalies because, after many years of mainstream
academia, he had come to feel that something important deep in his soul was longing for more meaningful
expression and that the study of anomalies offered an
opportunity to ask some profound questions about life
and consciousness. At that point in the conversation
Brenda pulled out an apple she had pocketed at lunch
and offered it to him, and they shared the "forbidden
fruit" just as the sun was setting and a full moon was
rising and the ambient light took on an almost mystical quality. We saved the seeds of the apple, which
were later secreted in the cornerstone of our new laboratory, and every year thereafter we returned to that
spot in the woods on the anniversary of that event.

On many of these visits we had what felt like magical encounters with deer who frequently approached us without fear. On one such occasion, a white deer emerged from the woods just a few feet away, and stood looking at us for several minutes. Another time, as we were engaged in deep conversation we looked up to discover that we were surrounded by eight adult deer, all watching us intently.

Carl Patrick

Deer were not the only creatures we encountered on these visits, although they were the most frequent. We also met foxes, raccoons, owls, flocks of geese, and several unusual species of birds, most of whom seemed to be aware of our presence and attempting to communicate with us in their own ways.

Such animal interactions also occurred frequently in other locations in the course of our travels. On a visit to Yellowstone Park, for instance, a white wolf emerged from the woods nearby and looked at us fixedly before silently disappearing back into the trees. And, as described in Chapter 15, we also had several remarkable encounters with cats.

But the animal that actually came to sym-bolize the PEAR program was the frog. It began on Brenda's first day in Princeton when she was feeling apprehensive about this new undertaking. Bob had picked her up to take her to the university and on the way had stopped briefly at a local pharmacy. While he was waiting for his prescription, Brenda wandered around the store and noticed some toy animals on a shelf, one of which was a small green frog with a goofy expression that made her smile and forget her worries. She purchased it and set this whimsical crea-ture on her new desk as a reminder to retain a sense of humor. A short time later, when we were taking some photographs of the laboratory for an invited article Bob was preparing for *The Proceedings of the IEEE*,[31] Brenda impulsively set the frog on top of the random event generator and it made its photographic debut in a refereed engineering journal. The popularity of that article resulted in a deluge of toy frogs and pictures of frogs from well-wishers all over the world and we soon accumulated a substantial collection in all shapes and sizes. Although we were not aware of it at the time, we later discovered that among its many associ-ations in different cultures, frogs were widely regarded

31 R. G. Jahn (1982). "The Persistent Paradox of Psychic Phenomena: An Engineering Perspective." *Proceedings of the IEEE*, 70 (2) pp. 136-170.

as good-luck tokens for gamblers, as well as being a symbol of abundance. They certainly served that purpose for PEAR. Many years later, when we developed a small mobile robot as an experimental device, we placed the original frog on the robot where it happily rode around its table and brought smiles to our many visitors—as well as success to our robot experiment.

PEAR frog with robot

18. Science and Spirit

ONE CANNOT pursue any serious study of anomalous physical phenomena very far without running headlong into spiritual issues. Although scientific education and social influences can readily raise doubts and questions regarding the veracity of canonical answers to the essential questions of the nature of life and reality, they cannot address these questions satisfactorily, much less resolve them. Nor can they accommodate the repeated empirical indications of the non-locality of consciousness, which inevitably raise questions about the reliability, or at least the completeness, of the conventional scientific representation of existence.

The two of us came to the study of anomalies from vastly different spiritual backgrounds and belief systems, but in the course of our work we both came to recognize the profundity of Albert Einstein's words, "Science without religion is lame; religion without science is blind,"[32] and our approach to the research at PEAR attempted to incorporate both of these dimensions in a complementary fashion. Our experiments were designed with scrupulous attention to the principles of the scientific method: protocols, data

32 A. Einstein (1941). "Science, Philosophy and Religion: a Symposium." A conference on Science, Philosophy and Religion in Their Relation to the Democratic Way of Life, New York.

processing, recording, analyses, and the reporting of results, all conformed to rigorous scientific standards. At the same time, the ambience of the laboratory, our interactions with our operators and visitors, and most importantly, our own attitudes toward each other and our work, all have reflected a deep respect for the spiritual dimensions of the topic. We have done our best to maintain an approach that has regarded the opportunity to pursue this program with an attitude of humility and with a sense of gratitude and privilege to be able to serve a lofty purpose, rather than seeking popularity or public or academic commendation.

As the empirical results accumulated, so did the evidence that we were looking at phenomena that transcended the standard physical dimensions of reality and challenged any simple materialistic worldview. Beyond its primary variable of intention, frequent reports from our operators suggested that successful results also required that they establish an emotional connection with the tasks at hand, a relationship we have come to refer to as "resonance." We also found that couples who were in love with each other produced significantly larger effects than either of them did when functioning as individuals,[33] and random event generator experiments conducted in group environments fostering relatively intense or profound subjective resonance showed larger anomalous deviations

33 B.J. Dunne (1991). "Cooperator Experiments with an REG Device." PEAR Technical Report #91005.

than those generated in more pragmatic assemblies.[34] And we often noted that when the dynamic interaction among the PEAR staff members was at its warmest and most collaborative, the experimental results appeared to reflect this resonance with stronger and more consistent results.

Nowhere was this interpersonal dimension more evident than in an experience we had with a physicist colleague who undertook an experiment involving a so-called "double-slit" apparatus, where participants attempted mentally to direct the flow of photons toward one of two openings.[35] When he was unable to demonstrate any anomalous effects, he suggested that he bring his equipment to us in Princeton, in case we would like to try the experiment ourselves. We accepted his proposal, and when he arrived we asked him describe his protocol and to show us how he introduced the experiment to his participants, whom he referred to as "subjects." This involved a fairly lengthy description of optics and quantum mechanics, followed by a statement to the effect that "some people actually believe that this process can be affected just by thinking about it." We recognized at once that this may have been the source of his problem and

34 R.D. Nelson et al (1998). "FieldREG II: Consciousness Field Effects." *Journal of Scientific Exploration*, Vol. 12, No. 3, pp. 425–454

35 M. Ibison and S. Jeffers (1998). "A Double-Slit Diffraction Experiment to Investigate Claims of Consciousness-Related Anomalies." *J. Scientific Exploration*, 12, No.4, pp.543–550.

decided to take a different approach with our "operators," simply informing them that we had a new device they were welcome to play with if they wished. They were told it was based on an optical process that we could explain in detail if they chose, but that they should simply try to make the indicator column on the computer display, which reflected the proportion of photons travelling through each slit, to move up or down. Several of our operators thought it would be fun to try and ended up producing a highly significant cumulative effect. All of this reinforced our sense that resonant interpersonal connections, including that of our own personal bond, have been major factors in the success of the PEAR experiments.

As we pondered the implications of this dimension for a deeper understanding of the nature of consciousness, we came to suspect that we were being given a glimpse of a primordial process that emanates from what is usually regarded as an unconscious domain of experience, one that is endemic to all living systems, and may even be the driving force of evolution. There is empirical evidence that some animals are capable of generating such anomalous effects under conditions of emotional involvement. For example, the French biol-

ogist Rene Peo'ch induced a group of baby chicks to imprint on a randomly driven robot, and when the chicks were separated by

a cage from the robot, which they regarded as their mother, the robot was found to spend a disproportionate amount of time near the chicks' cage.[36] The capacity of these animals to affect the trajectory of the robot to their biological advantage by some anomalous means lends credence to the hypothesis that we may be dealing with a phenomenon that is fundamentally biological in nature.

Just as we know that many biological processes operate at an unconscious level—breathing, cardiac activity, endocrine functions, etc.—we also know that our emotional experiences can affect our physiology as well as our minds. It appeared that these anomalous phenomena might be expressions of a deep unconscious biological capacity for organizing information. In other words, they might be evidence of the life force itself—what French philosopher Henri Bergson spoke of as the *élan vital* that underlies the creation of all living things, a process of self-organization that he linked closely with consciousness.[37, 38]

The history of science bears witness to the fact that virtually all scholars who developed and

---

36 R. Péoc'h (1988). "Chicken imprinting and the tychoscope: An Anpsi experiment." *Journal of the Society for Psychical Research*, 55, 1.

37 H. Bergson (1911). *Creative Evolution*, tr., Arthur Mitchell, New York: Dover, 1998.

38 B.J. Dunne and R.G. Jahn (2015). "Consciousness and the Nature of Life." Paper presented at the International Congress of Conscientology, Evora, Portugal, May 2015.

refined the scientific method had deep spiritual roots, and that their earliest investigations were motivated by metaphysical concerns. The early scientific heritage that evolved through the cultures of the Egyptians, Greeks, Romans, Orientals, Byzantines, and Medieval alchemists involved intimate admixtures of metaphysical rituals with rigorous analytical techniques, yet they generated extensive pragmatic knowledge and products, some of which, like the ancient pyramids or stone circles, still defy modern replication or full comprehension.

Marzolino/iStock

Even in the 20th century, the founders of relativity and quantum mechanics recognized and contemplated the subjective dimensions of reality.[39] Contemporary

Sir Isaac Newton, renowned scientist and alchemist

science still tends to ignore, or at least to minimize, this heritage and to dismiss such issues in favor of its preferred materialistic "exact" models. Yet, as science and its derivative technologies press forward into increasingly abstract and probabilistic domains, the relationship of spirit and consciousness—whether

39 R.G. Jahn and B.J. Dunne (2011). *Quirks of the Quantum Mind*. Princeton, NJ: The ICRL Press. Appendix.

divine or human, individual or collective—in the structure and operation of the physical world can no longer be casually set aside if the goal is a truly comprehensive understanding of nature.

19. Exodus

IN FEBRUARY of 2007, as the time approached to close the PEAR lab, we prepared a press release announcing that event.[40] Before issuing it, however, we were contacted by a reporter from the *New York Times* who had heard about the forthcoming closing and wanted to interview us for a feature article. We told him we would be issuing a press release in a few days, but he responded that if we agreed to his interview we would not need a press release, and so we agreed to speak with him. His article appeared on the front page of the *Times*' Science section on Saturday, February 10th, together with a large photograph of the two of us along with Percy, Bob's Labrador retriever, and Murphy, the large pinball machine that had dominated our reception area for so many years.[41]

As promised, this article negated the immediate need for an official press release, and by the following Monday we had been overwhelmed by countless representatives of the TV, radio, and print media from around the world. The announcement of our closing even appeared in the prestigious science journals, *Science* and *Nature*, neither of which had seen fit to mention our program throughout its 28 years of operation. Although the *Times* article had stated clearly

---

40 http://www.princeton.edu/~pear/press-statement.html
41 http://pearproposition.com/NYTarticle.html

that "at the end of the month, the Princeton Engineering Anomalies Research laboratory, or PEAR, will close, not because of controversy but because, its founder says, it is time,"[42] not because, as many of these representations implied, that the laboratory was being closed by the university due to the controversial nature of its studies. (The *Nature* article even went so far as to headline their article "The Lab That Asked the Wrong Questions."[43])

When we contacted the university's public relations office to issue our own press release, it was clear that they too had been barraged by media requests asking them to explain why our laboratory was being closed. We explained the situation and assured them that we had always made it clear that we felt privileged to have conducted our work at an institution that so honored the principle of academic freedom.

A few weeks later, several of our colleagues and former interns returned to Princeton to help us pack our files and equipment, and to join us in bidding farewell to the magical space that had been so meaningful to all of us. Each of them took home one of the stuffed animals that had adorned the laboratory and

---

42 B. Carey (2007). "A Princeton Lab on ESP Plans to Close Its Doors." *New York Times*, February 10, 2007. <http://www.nytimes.com/2007/02/10/science/10princeton. html?pagewanted=all&_r=0>

43 http://www.nature.com/nature/journal/v446/n7131/ full/446010a.html

contributed to its unique ambience, and in turn they presented us with two precious gifts. One was a hand-made, stained-glass pear; the other was a lovely book of photographs and personal tributes called *The PEAR Experience* that included a moving dedication express-ing their love and deep appreciation for the value that experience had had for their lives. It spoke of PEAR as a "second home....a cherished community where we found the inner strength to pursue our dreams,...and a haven to escape from the expectations of a judgmen-tal society," and expressed the wish that the spirit of PEAR would continue to thrive in the future.

We watched somewhat sadly as the physical evidence of the prior 28 years gradually disappeared into crates and boxes, and the various pieces of exper-imental equipment were directed into storage. It was especially moving when Murphy, the huge random

The PEAR Experience

mechanical cascade that had occupied an entire wall of the reception area and had become a widely recognized PEAR symbol, was removed. At that point, we were left with an empty space that continued to resonate with a treasury of precious memories, along with a sense of deep gratification that we had been able to accomplish so successfully what we had originally set out to do.

20. Metamorphosis

IN 1996, in anticipation of the eventual termination of the PEAR program at Princeton University, we had established a small not-for-profit organization that we called International Consciousness Research Laboratories, or ICRL. Its participants were drawn from an informal interdisciplinary consortium of professional colleagues we had organized several years earlier. We later expanded the group to include an assemblage of former PEAR colleagues and interns that we had whimsically labeled the "PEARtree." ICRL waxed as PEAR waned, and by 2007, when the PEAR laboratory closed, this extended family comprised some 80 participants from 15 countries, and featured a broad range of artistic and scholarly proficiencies.

Although the era of our formal laboratory research had come to a close in 2007, it was not clear at first exactly where our path would lead next. Given the limited resources at our disposal, it was necessary to consider the various options before us and determine which ones we could pursue most efficiently and effectively in order to continue the spiral snail-like expansion of the PEAR tradition within the ICRL context. Although we remained committed to furthering the scholarly exploration of consciousness-related anomalies, we felt that at this point we could be most effective as mentors and advisors to the next generation of researchers. Eventually, we finally decided to concentrate on the educational and communications components of our original mission.

Together with a small staff of dedicated volunteers, we undertook to organize a sequence of informational podcasts and host a series of stimulating speakers at local Meetup gatherings. And we established our own ICRL Press publishing imprint that would enable us to reissue *Margins of Reality* and to publish additional books that we regarded as pertinent to further explorations of the nature of consciousness. In the subsequent five years we have published a total of 10 books, some authored by ourselves and some by others, with additional manuscripts currently in various stages of preparation.

ICRL PRESS BOOKS

Now Available as Kindle eBooks

Amazon.com
Barnes&Noble.com
Google Books

And so our molecular journey continues. As the years have gone by and our bond has matured and strengthened, it seems as if our individual spirits have merged into one. The experience is closely akin to

Aristotle's profound observation: "Love is composed of a single soul inhabiting two bodies."[44]

As we reflect on our many shared challenges, accomplishments, adventures, and experiences, we take immense gratification in what we were privileged to achieve together. One important lesson we learned from our research is the realization that the accumulation of small effects can compound to a significant shift in the mean of a statistical distribution of random events. The implications of this modest statement apply not only to engineering anomalies but to any system that incorporates a degree of uncertainty—which includes virtually every aspect of reality from the microscopic to the macroscopic.

But our feelings of privilege extend well beyond any satisfaction in those academic dimensions. When we consider the broad community of colleagues, students, and friends that we have inspired and encouraged, we recognize that in them we have bequeathed a precious legacy. By responding to Einstein's dictum to maintain the complementarity of science and spirit, we may have propagated our most profound spiritual and scientific insight.

---

44 According to Laertius: To the query, "What is a friend?" Aristotle's reply was "A single soul dwelling in two bodies." (Diogenes Laërtius, *Lives and Opinions of Eminent Philosophers. Book 5: The Peripatetics*, "Aristotle," 9)

CPSIA information can be obtained
at www.ICGtesting.com
Printed in the USA
FSOW04n0440020417
32490FS